Waste and the Future
Sustainable Solutions from Earth, Space, and the Human Body

Harnessing Biological Efficiency for Waste Management on Earth and in Space

Foreword

As we journey further into the 21st century, the global challenges posed by waste continue to escalate. On Earth, waste impacts our environment, depletes our natural resources, and contributes to climate change. Meanwhile, as space exploration progresses, the issue of waste management in space missions becomes increasingly critical.

In this groundbreaking book, the authors delve into an often overlooked yet profoundly important source of inspiration for waste management: the human body. Nature has already provided us with the ultimate model for efficiency. Through years of evolution, the body has perfected waste management systems that filter toxins, repurpose energy, and eliminate waste with minimal loss.

"Waste and the Future" skillfully bridges the gap between biology, technology, and sustainability. By drawing lessons from the human body's processes, the authors present a visionary approach to rethinking waste on Earth and beyond. This book is not only a guide to the science behind sustainable waste management but also a call to action for a future where human ingenuity and nature's wisdom work in harmony.

Dr. Jenna Jambeck
Researcher

Preface

As we face mounting challenges with waste on our planet and prepare to explore new frontiers in space, the need for sustainable and efficient waste management systems has never been greater. Throughout history, humanity has struggled with waste—first from simple organic matter, and now from the complex, non-biodegradable waste of modern industry. Space exploration introduces a new dimension to this problem, where managing waste on long-term missions is crucial for survival.

This book explores an innovative approach to waste management by looking to nature's most efficient waste manager: the human body. By examining how the body filters, repurposes, and expels waste with remarkable efficiency, we can uncover models to help solve our waste crises both on Earth and in space. From recycling and energy recovery to creating zero-waste cities and closed-loop space habitats, this book offers a vision of a future where waste is minimized, resources are maximized, and sustainability becomes reality.

Dasbang, F. Joseph
Researcher/Author

Content

1. Foreword
2. Preface
3. Chapter 1: Waste on Earth: A Persistent Problem

- Overview of waste types: Organic, Inorganic, Hazardous, and Energy-Intensive
- Global waste production statistics and challenges
- The limitations of recycling and waste-to-energy solutions
- The environmental impact of waste on climate change and pollution

4. Chapter 2: Waste in Space: A Critical Frontier

- Types of waste generated in space missions
- Orbital debris and space junk concerns
- Challenges of waste management in confined space environments
- Innovations in space waste recycling, including NASA's efforts

Content

5.Chapter 3: The Human Body's Waste Management System

- The body's highly efficient processes: Digestion, Filtration, and Energy Conversion
- How the liver, kidneys, and metabolism manage waste
- Insights from human biology as a model for sustainable waste systems

6.Chapter 4: Lessons from the Body: Rethinking Waste on Earth

- Applying the body's selective filtration to industrial recycling
- Optimizing energy conversion inspired by metabolic processes
- Designing closed-loop systems for industries and cities

Content

7. Chapter 5: Waste in Space: Innovating with Biological Efficiency

- Adapting biological systems for space habitats
- Waste-to-energy technologies for self-sustaining space missions
- Developing fully closed-loop systems for future Mars and Moon colonies

8. Chapter 6: The Future of Waste Management: Earth and Beyond

- Vision for zero-waste cities on Earth
- Circular economies and sustainable industrial practices
- Designing sustainable space habitats with efficient waste systems
- The intersection of biological insight and technological innovation

Content

9. Conclusion: A Sustainable Future Inspired by Nature

- Summarizing the potential of biological efficiency in waste management
- The future of waste reduction, both on Earth and in space
- Final thoughts on creating a zero-waste, resource-efficient world

10. Bibliography

Chapter 1: Waste on Earth: A Persistent Problem

"Explores the vast amounts of waste produced globally, focusing on the challenges of managing organic, inorganic, hazardous, and energy-intensive waste. Highlights the limitations of recycling, the dangers of landfills, and the environmental impact of mismanaged waste on climate change."

Chapter 1: Waste on Earth: A Relentless Problem

Waste on Earth exists in many forms, each posing distinct challenges. Major types of wastes include organic, inorganic, hazardous, and energy-intensive wastes. Organic wastes, though easily biodegradable, often find an entrance into landfills, contributing to methane emissions—a very potent greenhouse gas. Inorganic wastes, mainly plastics and metals, are non-biodegradable and require vast resources for recycling. Hazardous waste, including medical and e-waste, contains chemicals that leach into the soil and water, causing long-term damage to ecosystems and human health. Energy-intensive wastes are complex materials such as multi-layer packaging, which are hard to process and recycle.

On a global scale, the amount of waste produced is staggering. According to the World Bank, about 2.01 billion tons of municipal solid waste are generated annually, and this is projected to increase to 3.40 billion tons by 2050. Of this, about 33% is not managed properly, leading to widespread pollution of oceans, rivers, and landscapes. A major issue is landfill capacity, which has been relied on for decades but is now reaching its limits. Additionally, improper waste disposal in oceans results in phenomena like the Great Pacific Garbage Patch, a massive accumulation of plastic debris.

Recycling, often cited as a solution to waste problems, is not a cure-all. While aluminum and paper are relatively easy to recycle, plastics pose a far greater challenge. Only about 9% of all plastic ever produced has been recycled, with the rest ending up in landfills or the environment. Furthermore, recycling itself is energy-intensive and often results in downcycling, where the quality of the recycled material is reduced with each cycle.

Waste-to-energy, where waste is incinerated to produce electricity, has been proposed as an alternative to landfilling. However, this method has its own drawbacks, such as the release of harmful chemicals like dioxins and furans into the atmosphere. The ash produced also contains toxic substances that require special disposal. Waste-to-energy should be seen as a stop-gap rather than a long-term solution.

Mismanaged waste contributes not only to local pollution but also significantly to global climate change. Organic waste in landfills produces methane, a greenhouse gas 28 times more potent than carbon dioxide over a 100-year period. Additionally, the production of new materials, especially plastics, depends heavily on fossil fuels, contributing to carbon emissions.

A comprehensive rethink of waste management is urgently needed—one that reduces waste at the source, optimizes recycling, and seeks innovative methods for energy recovery.

References: "Zero Waste Management Practices for Environmental Sustainability" Edited By Ashok Rathoure.

Zero Waste Home: "The Ultimate Guide to Simplifying Your Life by Reducing Your Waste" by Bea Johnson

Chapter 2: Waste in Space: A Critical Frontier

"Discusses waste generated in space missions, including biological and material waste, as well as the growing problem of orbital debris. Explores the unique challenges of waste management in space and the need for innovative recycling systems to sustain long-duration missions."

Chapter 2: Waste in Space: A Critical Frontier

The problem of waste is not unique to Earth. As mankind starts to venture into space, waste management is considered a critical issue in this new frontier. Examples of generated wastes from space missions are biological wastes, such as urine, feces, and food scraps; material wastes, such as packaging and broken equipment; and hazardous wastes, including toxic chemicals from experiments. Additionally, orbital debris accumulated from dead satellites, rocket parts, and other space junk has posed significant risks to active satellites and future space missions.

In space, waste management adds complexity. On Earth, waste can be transported to landfills or recycling facilities. In space, neither is an option. Everything taken aboard a spacecraft must be dealt with within the small confines of the vessel. NASA's International Space Station (ISS), for instance, uses a closed-loop system that treats, recycles, and filters water from urine to produce drinkable water. While this system is highly efficient, future missions to the Moon or Mars will require even more advanced waste recycling technologies.

Another growing issue is orbital debris, or space junk, which is increasingly becoming a major concern for space agencies worldwide. The European Space Agency estimates that over 36,500 objects larger than 10 cm are orbiting Earth, along with millions of smaller particles. These objects pose serious hazards to operational spacecraft; collisions can have catastrophic consequences. Space agencies are now exploring ways to address this problem, including using robotic arms or nets to remove debris from orbit.

In space, the stakes are higher due to limited energy resources. Traditional waste disposal methods, such as incineration, are not feasible because they consume too much energy. Space waste management systems must operate with minimal energy usage and maximize resource recovery. This has led to the development of innovative technologies like bioreactors, which convert organic waste into methane that can be used as fuel.

As space missions grow longer and more complex, waste management will become increasingly important. Future space habitats, such as those on Mars or the Moon, will require closed-loop systems that recycle nearly all waste into usable resources. Overcoming these waste challenges is crucial to achieving a sustainable human presence in space.

Reference: Space: "The Fragile Frontier" by Mark Williamson (Author)

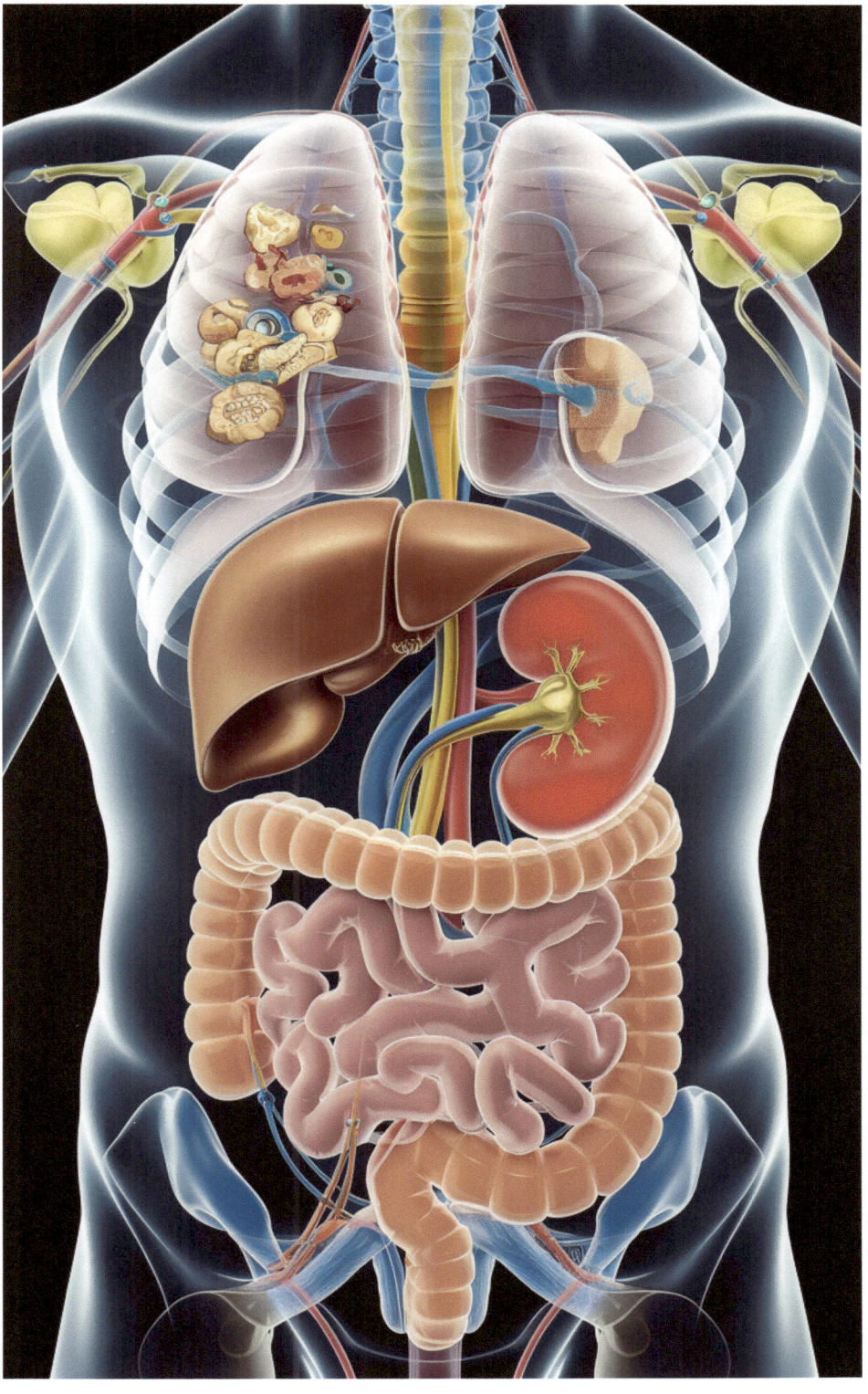

Chapter 3: The Human Body's Waste Management System

"Describes how the human body efficiently manages waste through digestion, filtration, and energy conversion. Shows how the liver and kidneys process and filter waste, providing a model for sustainable waste systems on Earth."

Chapter 3: Waste Management in the Human Body

The human body is an extremely efficient machine that has evolved over millions of years to manage its waste with minimal loss of valuable materials. In contrast to human-made systems for waste management, which are usually linear—wastes are generated and discarded—the human body operates within a closed system, continuously filtering, reworking, and eliminating waste products while reutilizing as much of the waste's components as possible.

Ingestion & Digestion: When food is ingested, it is broken down into key nutrients through digestion. The body absorbs what it needs for energy and cellular function, designating the rest as waste. This selective absorption ensures that nothing valuable is wasted.

Filtration & Separation: The liver and kidneys are the body's primary filtration systems. The liver breaks down poisons and other harmful substances into less dangerous compounds that can be safely excreted. The kidneys, meanwhile, filter blood, removing waste products like urea and creatinine while reabsorbing essential elements like water, electrolytes, and nutrients. Remarkably, kidneys filter about 180 liters of blood daily, retaining useful substances while producing only about two liters of urine.

Energy Conversion: Metabolism breaks down food into energy, extracting the maximum energy from nutrients with minimal waste. By-products such as carbon dioxide are efficiently expelled through respiration. This ensures the body operates with minimal energy loss.

Waste Expulsion: After filtering and processing, waste is expelled through the urinary and digestive systems. The separation of liquid and solid waste ensures efficient elimination, reducing the risk of infection or contamination.

The body's sophisticated waste management system can serve as a model for designing more effective waste systems on Earth. By mimicking these biological processes, we could develop technologies that optimize filtration, energy recovery, and waste separation.

Reference: *Guyton and Hall "Textbook of Medical Physiology" by John E. Hall, provides an in-depth explanation of how the human body processes waste and maintains internal balance.*

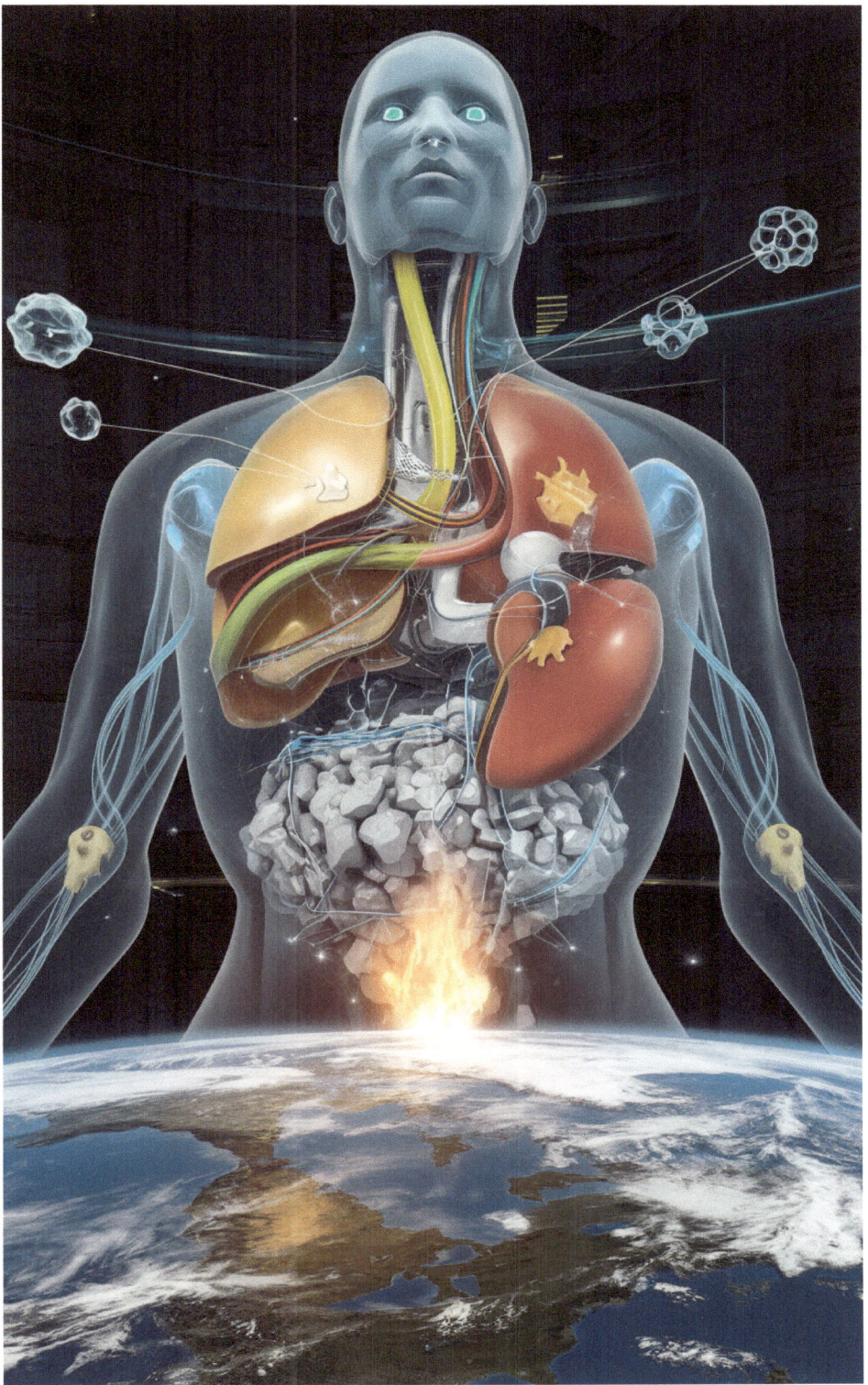

Chapter 4: Lessons from the Body: Rethinking Waste on Earth

"Applies biological waste management lessons to Earth's recycling challenges, emphasizing selective filtration, efficient energy conversion, and closed-loop systems to minimize waste and maximize resource recovery."

Chapter 4: Lessons from the Body: Reconsideration of Waste on Earth

The body's waste management system holds several critical lessons that can be applied to improve waste handling on Earth. By studying how the body separates, filters, and repurposes waste, we can design technologies that reduce environmental impact and enhance resource efficiency.

Selective Filtration (Inspired by the Liver and Kidneys): The liver and kidneys filter out toxins and waste while retaining valuable nutrients and water. On Earth, this could translate into advanced recycling systems. For example, future recycling plants could use sophisticated filtration technologies to separate valuable materials from waste streams. These systems could operate at the molecular level, ensuring that nothing valuable is lost.

Efficient Energy Conversion (Inspired by Metabolism): The body's metabolic processes convert food into energy with remarkable efficiency. Similarly, waste-to-energy technologies could be optimized to extract the maximum energy from waste materials. Bioreactors that convert organic waste into biofuels, for instance, can reduce reliance on fossil fuels. By mimicking metabolic pathways, these systems can operate with minimal energy loss, making them more sustainable and cost-effective.

Closed-Loop Systems (Inspired by the Body's Zero-Waste Approach): The body functions as a closed-loop system, where nearly every byproduct is repurposed or efficiently expelled. This idea can be applied to industrial waste management by creating closed-loop systems that recycle materials indefinitely. For example, companies like Terracycle are pioneering circular economy platforms, where products are designed to be fully recyclable, ensuring no material is wasted.

By drawing inspiration from the human body's waste systems, we can move toward a future where waste is minimized, resources are maximized, and environmental impact is reduced.

Reference: Stahel's "The Circular Economy: A User's Guide" offers practical insights into designing closed-loop systems that mimic nature's efficiency.

Chapter 5: Waste in Space: Innovating with Biological Efficiency

"Focuses on the need for closed-loop waste management systems in space, inspired by the human body's processes. Highlights how bioreactors and selective filtration technologies could make future space habitats sustainable."

Chapter 5: Waste in Space: Innovating with Biological Efficiency

As space exploration advances, the need for efficient waste management systems becomes more pressing. In space, waste can neither be carelessly discarded nor simply eliminated but must be recycled or repurposed to sustain long-duration missions. The human body's biological efficiency provides a model for developing these systems, ensuring that future space habitats generate minimal waste and maximize resource recovery.

Selective Filtration (Inspired by the Kidneys): NASA has already implemented selective filtration technologies aboard the International Space Station (ISS). For instance, the ISS uses urine processing systems that filter and purify water from urine, recycling it for drinking. This process could be refined further to filter other waste types, such as food scraps and biological materials, ensuring that every resource is reused.

Efficient Energy Conversion (Inspired by Metabolism): Future space missions could use waste-to-energy technologies that mimic the body's metabolic processes. Bioreactors, for example, could convert organic waste into methane, which can serve as a fuel source. Such systems would ensure that space habitats are self-sustaining, reducing the need for resupply missions from Earth.

Closed-Loop Systems (Inspired by the Body's Zero-Waste Approach): Long-term space missions, such as those to Mars, will require fully closed-loop waste management systems. These systems must recycle nearly all waste into usable resources, whether it be water, oxygen, or building materials. By mimicking the body's ability to repurpose waste, future space habitats could achieve near-total sustainability.

The biological efficiency of the human body offers a roadmap for designing space habitats that can sustain life for extended periods without external resupply. These innovations will be crucial as humanity embarks on longer and more ambitious space missions.

Reference: Bob Krone's "Beyond Earth: The Future of Humans in Space" Provides a foundation for space planners and anyone interested in human settlement in the solar system.

Chapter 6: The Future of Waste Management: Earth and Beyond

The future of waste management lies in creating systems that are as efficient and sustainable as the human body. As cities grow and generate more waste, we must rethink how we manage resources, moving from linear waste systems to circular economies. The lessons learned from the body's natural waste management processes will be crucial to achieving this transformation.

Zero-Waste Cities: Cities of the future could function like biological systems, continuously recycling waste into new resources. Advanced recycling plants, bioreactors, and selective filtration technologies could ensure waste is minimized and resources maximized. Cities such as Copenhagen and San Francisco are already pioneering zero-waste initiatives, offering a glimpse into what a sustainable future could look like.

Space Habitats: As humanity ventures beyond Earth, sustainable waste management will be essential for survival. Space habitats on the Moon or Mars will need closed-loop systems that recycle nearly all waste into usable resources. These habitats will be designed with biological efficiency in mind, ensuring that every resource is repurposed and nothing is wasted.

The future of waste management will be shaped by the integration of human ingenuity and biological inspiration to create systems that are efficient, sustainable, and capable of supporting life both on Earth and in space.

Reference: Frosch & Gallopoulos' seminal paper on "Industrial Ecology" (Scientific American, 1989) introduced the concept of using natural ecosystems as models for designing sustainable industrial systems.

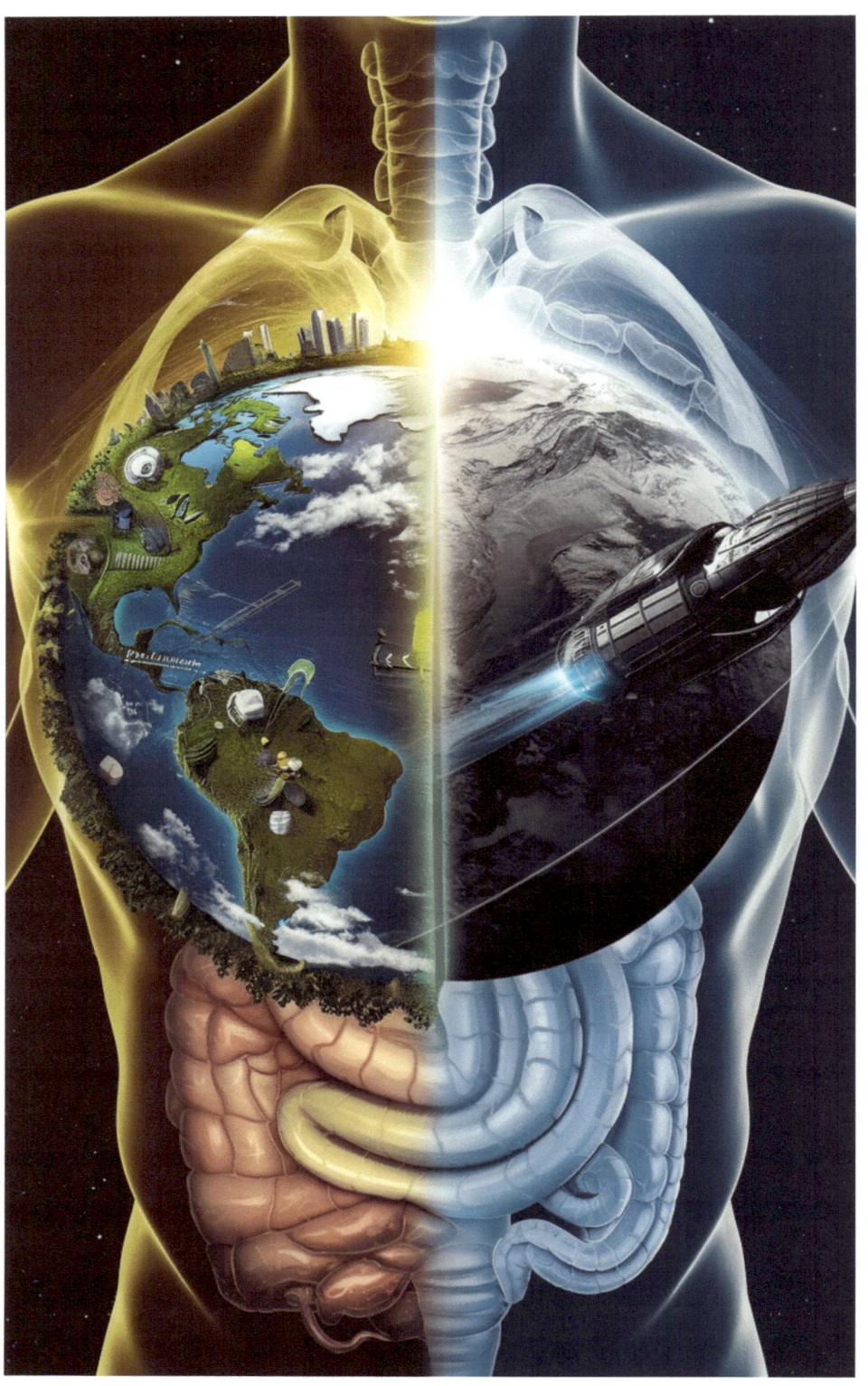

Conclusion: A Sustainable Future Inspired by Nature

By studying the human body's waste management systems, we can unlock new ways to address some of the most urgent environmental challenges of our time. From reducing waste on Earth to creating sustainable space habitats, biological efficiency offers a blueprint for a future where waste is minimized, resources are maximized, and life can thrive on Earth and beyond. This new, nature-inspired approach to waste management will be essential as we face the challenges of the 21st century and beyond.

Reference: Hawken, Lovins, and Lovins' "Natural Capitalism" explores how biological systems can inspire innovative solutions for sustainable development.

Bibliography

Hawken, Lovins, and Lovins' "Natural Capitalism"

Frosch & Gallopoulos' seminal paper on "Industrial Ecology" (Scientific American, 1989

"Zero Waste Management Practices for Environmental Sustainability" Edited By Ashok Rathoure.

Zero Waste Home: "The Ultimate Guide to Simplifying Your Life by Reducing Your Waste" by Bea Johnson

Space: "The Fragile Frontier" by Mark Williamson (Author)

Guyton and Hall "Textbook of Medical Physiology" by John E. Hall,

Stahel's "The Circular Economy: A User's Guide"

Bob Krone's "Beyond Earth: The Future of Humans in Space"

www.ingramcontent.com/pod-product-compliance
Lightning Source LLC
Chambersburg PA
CBHW040340220526
45473CB00009B/2746